Home Maths Ages 6–7

Anita Straker

CAMBRIDGE
UNIVERSITY PRESS

Ask an adult to time you.

You need pencil and paper. Write only the answers.

1   3 + 10.

2   20 – 1.

3   Write **fifty** in figures.

4   1 + 4 + 1.

5   10 take away 6.

6   2 × 4.

7   10 ÷ 5.

8   What is next: 6, 8, 10, 12 …?

9   Add 10 to 17.

10   4 plus 3.

11   Which 2 coins make 7p?

12   10 – □ = 8.

2

First to 20

Play with one or two others.

Share a pencil and paper.

Start with 0.

Take turns to add on any number from 1 to 5.

Write the new total on the paper.

The first to reach **exactly** 20 scores a point.

The winner is the first to get 10 points.

> **Change the rules**
>
> Start at 40.
>
> Take turns to subtract any number from 1 to 5.
>
> The first to reach exactly 20 scores a point.

Add small numbers

Lena's toys

Do this on your own.

You need pencil and paper.

Lena has 20p to buy 3 different toys.

She may get change, or she may not.

There are 6 different ways she can buy 3 toys.

Write a list of them.

By each set of 3 toys, write how much they cost altogether.

Now work out her change each time.

Find a total of three items and change from 20p
Work systematically

Place your order

Two, three or four people can play.

You need pencil, paper and scissors and a pack of playing cards.

Each player should use the ace to 10 of one suit.

Each of you should draw a grid like this.

The spaces should be big enough for a card to go in them.

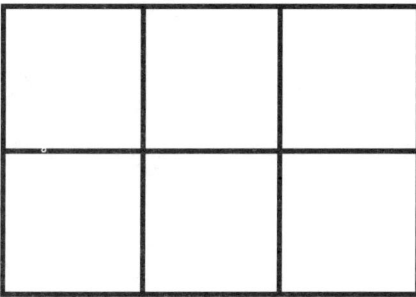

Turn your cards face down and shuffle them.

Take turns to look at one of your cards and put it on your grid.

Once you have placed it you must not move it.

The aim is to get the 3 numbers in each row in order, smallest first.

Score one point for each row that is in order.

Play several times.

The winner is the first
to get 10 points.

Change the rules

a. Draw a 3 by 3 grid.

b. Draw a 4 by 2 grid.

c. Make cards from 1 to 15
     for each player.

Put numbers in order
Make predictions

Ask an adult to read you these.
You need pencil and paper. Write only the answers.

1 Three fives.

2 Write **eighty** in figures.

3 Divide 10 by 5.

4 Take 7 from 10.

5 Double 3.

6 2 less than 7.

7 What is the total of 10p and 10p?

8 Half of 10.

9 Write two numbers that add up to 11.

10 How much altogether is 6p and 4p?

11 Multiply 10 by 4.

12 Write the difference between 8 and 5.

6

Bean race

Play with a partner.

You need a dice for each player and a pile of dried beans.

Say 'Go'.

Each of you roll your dice.

Double your number.

Take that number of beans.

Stop after five rolls.

Count your beans in twos.

The player who has closest to 40 beans scores a point.

The first to get 5 points wins.

Change the rules

a. The last to collect his or her beans puts 3 of them back.

b. Count the beans in fives.

c. Use 2 dice each and see who is closest to 70.

ouble numbers from 1 to 6
ount in twos or fives

7

Brian's stamps

Do this on your own.

You need pencil and paper.

Brian bought some 20p stamps and some 10p stamps.

He spent £2 altogether.

He bought three times as many 10p stamps as 20p stamps.

How many of each stamp did he buy?

How many 30p stamps can Brian buy with £2?

How much change would he get?

Add multiples of 10p and 20p
Work out a strategy

8

Ask an adult to time you.

You need pencil and paper. Write only the answers.

1 3 times 3.

2 14 – 4.

3 20 ÷ 10.

4 $3 + \square = 10$.

5 7 minus 0.

6 2×6.

7 Find the total of 4 kg, 5 kg and 1 kg.

8 29 – 10.

9 Which 3 coins make 6p?

10 Write **fifteen** in figures.

11 3 + 2 + 1.

12 What is next: 15, 20, 25, 30 …?

Worms

Two, three or four people can play.

You need a dice.

Each player needs pencil and paper.

Each player should draw a worm like this.

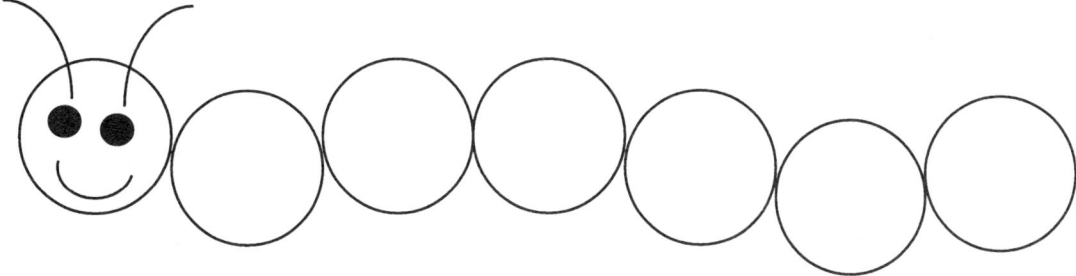

Take turns to roll the dice.

Double your score.

Write your answer in a space on your worm.

Carry on until your worm is full.

Now swap your worm with someone else.

Take turns to roll the dice again.

If your score is half a number on your new worm, cross out that number.

Otherwise wait for your next turn.

The first to cross out all their numbers wins.

Change the rules

a. Use two dice and double your total.

b. Use one dice and multiply by 10 or 5.

Double and halve small numbers

Pop the balloon

Play with a partner.

You need a dice.

You each need pencil and paper.

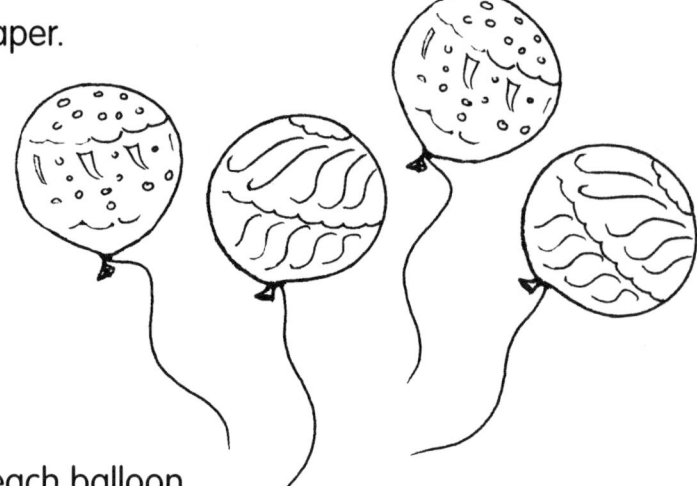

Draw four ballons each.

Write a different number on each balloon.

Choose from these numbers.

Take turns to roll the dice.

See if you can add one of your partner's balloon numbers to your dice number to make 10.

If so, cross out that number and 'pop' the balloon.

The player whose four balloons are popped first loses.

Play several times.

Change the rules

Choose from 14, 15, 16, 17, 18, 19.

Try to make 20.

Identify pairs of numbers with a sum of 10 or 20

Give-away

Three or four people can play.

You need 20 dried beans for each player, and two dice.

Take turns to roll both dice and say the total.

If you have enough, give that number of beans to the player on your left.

If not, wait for your next turn.

Pass the dice to the player on your right.

The winner is the first to get rid of all their beans.

Change the rules

Start with 10 beans each. Give away the difference between the numbers.

Add or subtract pairs of small numbers

Ask an adult to read you these.
You need pencil and paper. Write only the answers.

1 Two sixes.

2 4 more than 5.

3 Write the sum of 14 and 2.

4 Subtract 9 from 10.

5 Share 8 equally among 2.

6 9 minus 2.

7 Divide 6 by 3.

8 Which is more: 43 or 34?

9 Add 10 to 6.

10 Today is Sunday. What is tomorrow?

11 10 multiplied by 3.

12 Half of 16.

Ask an adult to time you.

You need pencil and paper. Write only the answers.

1   24 + 10.

2 2×3.

3   8 plus 0.

4 $\square$ + 2 = 7.

5   12 – 3.

6   2 + 4 + 2.

7   How many 10p coins in £1?

8   2 footballs at £7 each cost …?

9   13 is more than 9. How many more?

10   Multiply 5 by 3.

11   36 – 10.

12   What is next: 13, 15, 17, 19 …?

14

Add ups

Do this on your own.

You need pencil and paper.

Choose two different numbers from this set.

Add them.

There are 10 ways to choose and 9 different answers.

Can you list them all?

Change the rules

The two numbers must be the same.

What are the different answers now?

Add pairs of small numbers
Work systematically

Merlin's Castle

Play with a partner.

You need: pencil and paper to draw the track

a different button or coin for each player

a pack of playing cards (Use the ace to 10 of each suit.)

Draw a track to Merlin's Castle.

Put your buttons at the start.

Start

Shuffle the cards.

Put them face down in a pile.

Take turns to take the top two cards.

If the sum is odd, move your button forward one space.

If the sum is even, stay where you are.

If you use all the cards, shuffle them and start again.

The first to reach Merlin's Castle wins.

Change the rules

Spread the cards face down.

Take turns to turn over a pair.

Add two numbers with a total up to 20
Recognise odd and even numbers up to 20

Omar and his camel

Do this on your own.

You need pencil and paper.

Omar rides a camel.

His camel has a bag on each side.

These boxes of dates must go in the bags.

The two bags must balance each other.

How should Omar put the dates in the bags? Find two different ways.

Add small numbers
Work out a strategy

Ask an adult to read you these.

You need pencil and paper. Write only the answers.

1	Take 4 from 17.	**7**	Double 6.
2	One half of 2.	**8**	How much altogether is 5p and 8p?
3	Four tens.	**9**	What is the total of 5, 5 and 4?
4	3 less than 18.	**10**	Halve 14.
5	How many threes in 9?	**11**	Write the difference between 17 and 15.
6	6 plus 4.	**12**	Today is Friday. What was yesterday?

Party

Do this with a partner.

P A R T Y

1   2   3   4   5

The numbers stand for letters.

Add up the numbers in the words below.

Your partner should check and say if you are right.

1	AT	**5**	ART	**9**	PAY
2	TAP	**6**	RAY	**10**	TAR
3	RAT	**7**	TRY	**11**	PART
4	YAP	**8**	RAP	**12**	TRAY

Add several small numbers
Use addition facts to 20

Sum up

Do this by yourself.

You need pencil and paper.

Mrs Jones made up some sums for her class to do in their heads.

She used only the digits 2, 3 and 4.

For each sum, she used each digit once.

The sums are all like this.

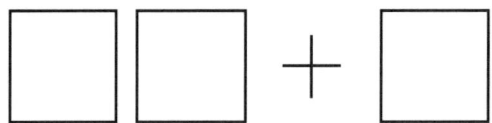

Altogether she has made up 6 different sums.

Between them, the sums have three different answers.

What are they?

Change the rules

This time, subtract the numbers.

☐☐ − ☐

What are the different answers now?

Kate's parcel

Do this on your own.

You need pencil and paper.

Kate wants to put 30p more in stamps on her parcel.

She has lots of 4p, 5p and 10p stamps.

She can do it in 6 different ways.

Can you write them all?

Add 10s, 5s and 4s
Work systematically

21

Ask an adult to time you.

You need pencil and paper. Write only the answers.

1	Double 11.	**7**	How many centimetres in 1 metre?
2	2 + 4 + 3.	**8**	2×7.
3	19 − 9.	**9**	How many ears have 5 cats?
4	5 times 5.	**10**	2 cakes at 40p each cost …?
5	$\square - 6 = 3$.	**11**	How many 5p coins in £1?
6	$40 \div 10$.	**12**	56 − 10.

Princess Emelia

Do this by yourself.

You need pencil and paper.

You need 20 'jewels'.

(Dried beans or buttons will do!)

Princess Emelia sorted her 20 jewels in piles.

She made some piles with 7 jewels in each.

All her other piles had 2 jewels each.

How many piles did she make?

This time Emelia sorted her 20 jewels in four piles.

The first pile had 4 more than the second.

The second pile had 1 less than the third.

The fourth pile had twice as many as the second.

How many jewels did she put in each pile?

Calculate with numbers up to 20
Work out a strategy

Home Maths Ages 6–7 Answers

1

1	13
2	19
3	50
4	6
5	4
6	8
7	2
8	14
9	27
10	7
11	5p, 2p
12	2

2

A player who must add to 2, 8, 14 … should lose, if the other player plays correctly. Whatever is added to 14, for example, will not be enough to make 20; the other player can then win on their next move.

Change the rules
Similarly, a player on 38, 32, 26 … should lose.

3

	cost	change
book, ball, top	20p	no
book, pen, top	18p	2p
puzzle, ball, pen	20p	no
puzzle, ball, top	18p	2p
puzzle, pen, top	16p	4p
ball, pen, top	15p	5p

Children could extend this puzzle by inventing their own set of priced toys.

5

1	15
2	80
3	2
4	3
5	6
6	5
7	20p
8	5
9	Any pair with a sum of 11
10	10p
11	40
12	3

7

Brian bought:
four 20p stamps
twelve 10p stamps

Brian can buy six 30p stamps with £2.
He will get 20p change.

8

1	9
2	10
3	2
4	7
5	7
6	12
7	10 kilograms (or 10 kg)
8	19
9	2p, 2p, 2p
10	15
11	6
12	35

12

1	12
2	9
3	16
4	1
5	4
6	7
7	2
8	43 is more
9	16
10	Monday
11	30
12	8

13

1	34
2	6
3	8
4	5
5	9
6	8
7	10
8	£14
9	4
10	15
11	26
12	21

14

$2 + 3 = 5$ $3 + 5 = 8$
$2 + 4 = 6$ $3 + 10 = 13$
$2 + 5 = 7$ $4 + 5 = 9$
$2 + 10 = 12$ $4 + 10 = 14$
$3 + 4 = 7$ $5 + 10 = 15$

Change the rules
$2 + 2 = 4$
$3 + 3 = 6$
$4 + 4 = 8$
$5 + 5 = 10$
$10 + 10 = 20$

16

Omar's camel bags should be packed with 13kg, or half the total weight, each side.

8kg and 5kg
6kg, 4kg, 2kg and 1kg

or

8kg, 4kg and 1kg
6kg, 5kg and 2kg

17

1	13
2	1
3	40
4	15
5	3
6	10
7	12
8	13p
9	14
10	7
11	2
12	Thursday

18

1	AT	5
2	TAP	7
3	RAT	9
4	YAP	8
5	ART	9
6	RAY	10
7	TRY	12
8	RAP	6
9	PAY	8
10	TAR	9
11	PART	10
12	TRAY	14

19

$43 + 2 = 42 + 3 = 45$
$32 + 4 = 34 + 2 = 36$
$24 + 3 = 23 + 4 = 27$

$43 - 2 = 41$
$42 - 3 = 39$
$34 - 2 = 32$
$32 - 4 = 28$
$24 - 3 = 21$
$23 - 4 = 19$

20

Stamps for Kate's parcel:

10p, 10p, 10p
10p, 10p, 5p, 5p
10p, 5p, 5p, 5p, 5p
10p, 4p, 4p, 4p, 4p, 4p
5p, 5p, 4p, 4p, 4p, 4p, 4p
5p, 5p, 5p, 5p, 5p, 5p

21

1	22
2	9
3	10
4	25
5	9
6	4
7	100 centimetres
8	14
9	10 ears
10	80p
11	20
12	46

22

Princess Emelia's piles of jewels were:

2 piles of 7 jewels
3 piles of 2 jewels

1st pile	7 jewels
2nd pile	3 jewels
3rd pile	4 jewels
4th pile	6 jewels

23

1	5
2	25
3	2 metres (or 2m)
4	17
5	18
6	22
7	106
8	8
9	14
10	89 is less
11	100
12	7 pairs

24

Total of 12:

Goo
Zup, Sig
Jid, Naf
Paz, Dib
Bud, Red
Vik, Tug
Zup, Jid, Dib

Zup, Paz, Red
Zup, Vik, Art
Zup, Bud, Tug
Jid, Paz, Tug
Jid, Bud, Art
Paz, Bud, Vik
Zup, Jid, Paz, Art
Zup, Bud, Jid, Vik

Change the rules
Make 14: 19 different ways
Make 16: 27 different ways

25

1	18 kilograms (or 18 kg)
2	70
3	10
4	17
5	15 is odd
6	14
7	20p, 2p, 2p
8	39
9	4
10	4
11	£15
12	2

27

Possibilities are	cost
jelly, jelly	4p
jelly, fudge	5p
fudge, fudge	6p
jelly, laces	7p
fudge, laces	8p
jelly, gum	9p
fudge, gum	10p
laces, laces	10p
jelly, gob stopper	11p
fudge, gob stopper	12p
laces, gum	12p

Least change is no change!

29

1	16
2	4
3	30
4	30
5	12
6	12
7	25p
8	21
9	10
10	18p
11	3
12	21

30

2	two 10p
3	10p, two 5p
4	four 5p
5	10p, 5p, two 2p, 1p
6	10p, five 2p
7	10p, 5p, five 1p
8	10p, three 2p, four 1p
9	10p, two 2p, six 1p
10	ten 2p
11	10p, ten 1p
12	two 5p, ten 1p
13	5p, three 2p, nine 1p
14	5p, two 2p, eleven 1p
15	5p, 2p, thirteen 1p

31

Eight on each side
```
3 2 3    4 2 2
2   2    2   2
3 2 3    2 2 4
```
Nine on each side
```
4 1 4    6 1 2
1   1    1   1
4 1 4    2 1 6
```
Six on each side
```
1 4 1
4   4
1 4 1
```
Seven on each side
```
1 3 3    2 3 2
3   3    3   3
3 3 1    2 3 2
```

32

1	20
2	28 is even
3	10
4	18 centimetres (or 18 cm)
5	80
6	9
7	20p, 2p, 2p, 1p
8	16
9	33p
10	2
11	30
12	14

35

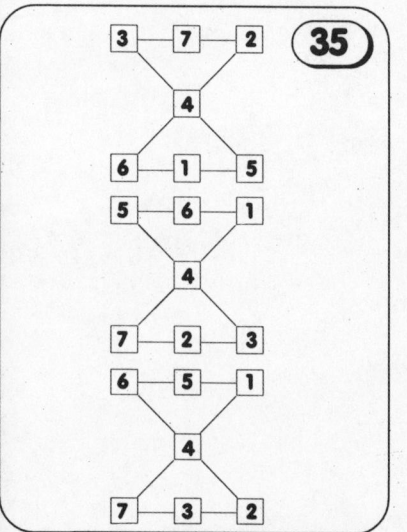

36

1	14
2	0
3	90
4	11
5	9
6	10
7	16
8	110
9	16 shoes
10	19p
11	2
12	1

37

1	16
2	100
3	18
4	13
5	3
6	40
7	eg, 21, 23, 25, 27 …
8	35 metres
9	44p
10	16 legs
11	6
12	4p

38

2	5	4	3	1	7	20	11	12	15	9	8
4	10	8	6	2	14	40	22	24	30	18	16
8	20	16	12	4	28	80	44	48	60	36	32

39

For example,

WOLF	23 + 6 = 29
OSTRICH	20 + 9 = 29
MONKEY	25 + 5 = 30
OWL	23 + 12 = 35
CAMEL	13 + 1 = 14
LAMB	13 + 1 = 14
LLAMA	13 + 1 = 14
CALF	12 + 1 = 13

41

1	8
2	12
3	18
4	7
5	15
6	100
7	13p
8	3
9	19
10	3
11	8
12	14 days

43

For example,

1	= 5 − 4 + 3 − 2 − 1
2	= 13 − 2 − 4 − 5
3	= 5 − 4 + 3 − 2 + 1
4	= 14 − 2 − 3 − 5
5	= 5 − 4 + 3 + 2 − 1
6	= 15 − 2 − 3 − 4
7	= 5 − 4 + 3 + 2 + 1
8	= 12 − 3 + 4 − 5
9	= 5 + 4 + 3 − 2 − 1
10	= 12 − 3 − 4 + 5

11	= 5 + 4 + 3 − 2 + 1
12	= 13 − 2 − 4 + 5
13	= 5 + 4 + 3 + 2 − 1
14	= 12 + 3 + 4 − 5
15	= 5 + 4 + 3 + 2 + 1
16	= 12 + 3 − 4 + 5
17	= 21 − 3 + 4 − 5
18	= 12 − 3 + 4 + 5
19	= 21 − 3 − 4 + 5
20	= 13 − 2 + 4 + 5

Can children go further?

44

1	40
2	8
3	12
4	12 kilograms (or 12 kg)
5	6
6	48
7	eg, 18, 16, 14, 12 …
8	£66
9	4
10	20
11	10
12	50p, 20p, 10p

Ask an adult to read you these.

You need pencil and paper. Write only the answers.

1 Subtract 6 from 11.

2 Five fives.

3 One quarter of 8 metres.

4 Write the sum of 10 and 7.

5 21 minus 3.

6 4 more than 18.

7 Write **one hundred and six** in figures.

8 Divide 16 by 2.

9 Add 3 to 11.

10 Which is less: 89 or 98?

11 Multiply 10 by 10.

12 14 socks make how many pairs?

24

Space men

Two, three or four people can play.

Share a pencil and paper.

You need some beans.

Take turns.

Write the names of the space men to go on the space ship.

They must be worth a total of 12.

If it is a new way, you win a bean.

Stop when no more ways can be found.

The winner has the most beans.

Change the rules

The space men must be worth a total of 14, or a total of 16.

Add numbers from 1 to 12
Record systematically

Ask an adult to time you.

You need pencil and paper. Write only the answers.

1   Add 9 kg to 9 kg.

2   50 + 20.

3   2 × 5.

4   20 − 3.

5   Is 15 odd or even?

6   4 less than 18.

7   Which 3 coins make 24p?

8   ☐ + 1 = 40.

9   How many 5p coins in 20p?

10   20 ÷ 5.

11   3 books at £5 each cost …?

12   51 − 49.

26

Gold coins

Play with a partner.

You are both pirates.

You need a pile of 21 'gold coins'.
(Dried beans or buttons will do!)

Take turns.

Take 1, 2 or 3 gold coins from the pile.

Carry on until all the gold coins are taken.

The pirate who has taken an **odd number** of gold coins scores 2 points.

Play several times.

The winner is the first to get 10 points.

Recognise odd and even numbers
Work out a strategy

Sweet shop

Play with a partner.

You each need pencil and paper.

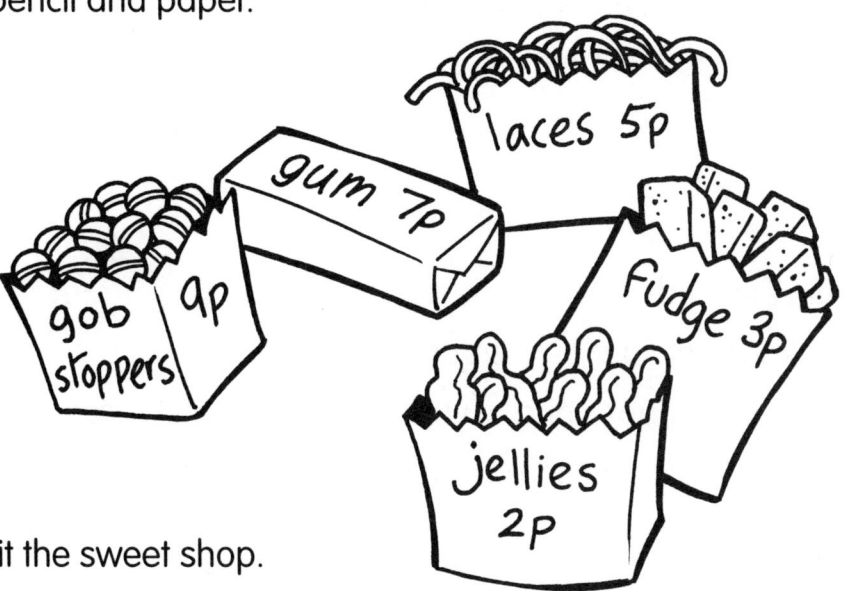

Take turns to visit the sweet shop.

Imagine you have 12p to spend.

What two sweets can you buy?

Write down your change.

Next turn you have 11p to spend,

then 10p, then 9p, then 8p, then 7p, then 6p.

Add up your change.

The winner is the one with the least change.

Change the rules

What three sweets can you buy?

Find a total cost and work out change
Think logically

It's tricky!

Two people can play.

You need a pack of playing cards.

Use the ace to 9 of both red suits.

Shuffle the cards. Deal 10 cards each.

Secretly look at your own cards.

Take turns to choose one of your cards and put it face up in the middle.

See if your card, and some or all of the others in the middle, add up to 15.

> It doesn't have to be all the cards that add up to 15.
>
> **Just some will do.**

If they do, take those cards - you have won a trick – but leave the other cards in the middle!

If they don't, leave your card in the middle and wait for your next turn.

Carry on until you both have no cards left.

The winner is the player to get the most tricks.

Play several times. Take turns to go first.

Change the rules

a.   Use three suits and deal 15 cards each.

b.   Make tricks worth 20.

Add small numbers

Ask an adult to read you these.

You need pencil and paper. Write only the answers.

1 Eight twos.

2 Take 8 from 12.

3 Multiply 3 by 10.

4 Double 15.

5 6 plus 6.

6 Subtract 3 from 15.

7 What is 10p more than 15p?

8 25 take away 4.

9 Add together 3, 5 and 2.

10 What is the total of nine 2p coins?

11 12 divided by 4.

12 What is the odd number after 19?

30

Toffees

Ask all your family to join in.

You need pencil and paper
and lots of 1p, 2p, 5p and 10p coins.

A bag of toffees costs 20p.

You can pay for it exactly using a 20p coin.

Write other ways of paying 20p exactly.

Can you do it using 2 coins? 3 coins? 4 coins?

Keep going.

Can you do it all the way up to using 15 coins?

Add coins to a total of 20p
Work systematically

Saucers

Do this on your own.

You need 8 saucers and 20 dried beans (or buttons).

Arrange the 8 saucers in a square.

Put the 20 beans on the saucers.

Put at least one on each saucer.

Get 8 beans on each side of the square.

Change the rules

a.   Make 9 on each side.

b.   Make 6 on each side.

c.   Make 7 on each side.

Adding three small numbers
Think logically to eliminate what won't work

32

Ask an adult to time you.
You need pencil and paper. Write only the answers.

1   15 add 5.

2   Is 28 odd or even?

3   1 + 4 + 5.

4   Take 2cm from 20cm.

5   100 − 20.

6   90 ÷ 10.

7   Which 4 coins make 25p?

8 2×8.

9   3 pens at 11p each cost …?

10   How many more than 19 is 21?

11 $\square - 1 = 29$.

12   What is next: 2, 5, 8, 11, …?

Unlucky for some

Two to four people can play.

You need pencil and paper for each player
and two dice.

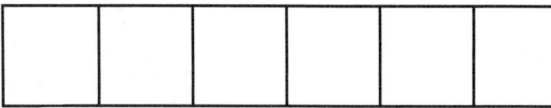

Each player should draw a grid like this.

Choose from the numbers 1 to 11.

Write a different number in each grid space.

Take turns.

Roll the two dice and add the numbers.

Work out how many more to make 13.

If you can see that number on another player's grid, you can cross it out.

Otherwise, wait for your next turn.

If all your numbers are crossed out, you are 'out'.

Keep going.

The winner is the last player with one or more numbers left.

> **Change the rules**
>
> Choose from the numbers 9 to 19 to write on the grid.
>
> Work out how many more to make 21.

Add pairs of small numbers
Find a difference from 13 or 21.

Beat the clock

Play with a partner who has a watch or clock with a second hand.

You need a pack of playing cards.

Use the ace to 10 of each red suit.

Shuffle the cards.

Put them face down in a pile.

Your partner should say 'Go' and time you.

Turn over the top card.

Take the number from 15.

If your partner says you are right, keep the card.

If not, put the card back at the bottom of the pile.

Stop after 1 minute.

How many cards did you win?

Play again.

This time carry on until you have won all the cards.

When you win the last card say 'Stop'.

How many minutes and seconds did you take?

Play again.

Can you beat your record?

Change the rules

a.   Take the number from 20.

b.   Take the number from 13.

Subtract a number from 15, 20 or 13

Dozen it

Do this on your own.

You need pencil, paper and some scissors.

Make 7 number cards like this.    **7**

Arrange the cards like this.

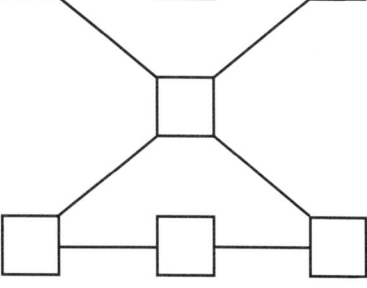

Each line of three numbers must add up to 12.

Can you find another way to do it?

Adding three small numbers
Think logically to eliminate what won't work

Ask an adult to read you these.
You need pencil and paper. Write only the answers.

1   Seven twos.

2   8 minus 8.

3   9 multiplied by 10.

4   6 more than 5.

5   Divide 18 by 2.

6   Add 2 to 8.

7   What is the even number before 18?

8   Write **one hundred and ten** in figures.

9   How many shoes in 8 pairs?

10   How much altogether is 12p and 7p?

11   Write the difference between 19 and 17.

12   One quarter of 4.

Ask an adult to time you.

You need pencil and paper. Write only the answers.

1 20 – 4.

2 70 + 30.

3 2 × 9.

4 3 + 6 + 4.

5 15 ÷ 5.

6 ☐ – 2 = 38.

7 Write 2 odd numbers more than 20.

8 Add 5 metres to 30 metres.

9 2 ices at 22p each cost …?

10 How many legs on 4 dogs?

11 ½ of 12.

12 How much more than 16p is 20p?

Halves and doubles

Do this on your own.

You need pencil and paper.

The top number is half the middle number.

The bottom number is double the middle number.

Copy this table.

Fill in the empty boxes.

2		**4**			**7**			**12**			
4	**10**			**2**		**40**				**18**	
8			**12**				**44**		**60**		**32**

Find halves and doubles in the range 1 to 60

Animals

Do this on your own.

You need pencil and paper.

Each letter stands for a number.

A	B	C	D	E	F	G	H	I	J	K	L	M	N	O	P	Q	R	S	T	U	V	W	X	Y	Z
1	2	3	4	5	6	7	8	9	10	11	12	13	14	15	16	17	18	19	20	21	22	23	24	25	26

Think of an animal.

Write its name on your paper.

Under each letter write its number.

Add the biggest number to the smallest number.

Write the answer.

Now try some more animals.

What is the biggest answer you can get? What is the smallest answer?

Change the rules

a. Instead of adding the numbers, find their difference.

b. Try colours, flowers, vegetables, or forms of transport …

Add or subtract a pair of numbers in the range 1 to 26
Work out a strategy

Strings

Play with a partner.

You need about 2 to 3 metres of string and some scissors, and a ruler marked in centimetres.

Cut the string into about 10 different lengths.

Make some shorter, some longer.

Your partner should hold up the strings one at a time.

You must say what you think the length is.

Now measure the length to the nearest centimetre.

Work out the difference from your estimate and find your score.

Difference	Score
10 cm or more	1 point
4 to 9 cm	2 points
1 to 3 cm	5 points
Spot on	10 points

Carry on until you have done each string.

If your total score is 50 points or more, you are a star!

Can someone in your family do better than you?

Estimate and measure lengths in centimetres

Ask an adult to read you these.

You need pencil and paper. Write only the answers.

1　4 less than 12.

2　3 plus 9.

3　Nine twos.

4　Divide 70 by 10.

5　Subtract 5 from 20.

6　Double 50.

7　What is the total of 2p, 4p and 7p?

8　How many fours in 12?

9　Write the sum of 14 and 5.

10　19 minus 16.

11　Take 5 from 13.

12　How many days in 2 weeks?

42

Do this on your own.

You need a pack of playing cards.

Use the ace to 10 of each suit.

Spread the 40 cards face up.

Make 17

7 + 3 + 7 = 17

Set out the cards in rows.

You can have as many cards as you like in each row.

Score 5 points for each row that adds up to 17.

How many points can you score?

Play again.

Can you beat your previous score?

Change the rules

Make rows that total 16, 18 or 19.

Add several numbers to total 17 (or 16, 18 or 19)

Fish alive

Do this on your own.

You need pencil and paper.

Make 'sums' using + and –.

Use only the digits 1, 2, 3, 4 and 5.

For each 'sum', use each digit once.

Try to get different answers. For example,

$$14 + 2 - 5 - 3 = 8$$

Can you get each of the numbers 1 to 20 as an answer?

Calculate with small numbers
Record systematically

Ask an adult to time you.

You need pencil and paper. Write only the answers.

1 90 – 50.

2 7 + 4 – 3.

3 ½ of 24.

4 Take 8kg from 20kg.

5 60 ÷ 10.

6 2 + ☐ = 50.

7 Write two even numbers less than 20.

8 2 coats at £33 each cost …?

9 How many more than 12 is 16?

10 12 + 8.

11 ☐ × 10 = 100.

12 Which 3 coins make 80p?

Zigzag

Play with a partner.

You need paper, a ruler marked in centimetres and a dice.

Each player needs a different coloured pen.

Use your ruler to draw a big zigzag, with each length
an exact number of centimetres.

Start

Home

Take turns to roll the two dice.

Add the numbers to find the total.

Measure the total number of centimetres along the track.

Make a mark with your pen.

Each time measure on from your last mark.

The winner is the first to reach home.

Draw another track and play again.

Use a ruler to measure in centimetres
Calculate how much further to measure

PUBLISHED BY THE PRESS SYNDICATE OF THE UNIVERSITY OF CAMBRIDGE
The Pitt Building, Trumpington Street, Cambridge CB2 1RP, United Kingdom

CAMBRIDGE UNIVERSITY PRESS
The Edinburgh Building, Cambridge CB2 2RU, United Kingdom
40 West 20th Street, New York, NY 10011-4211, USA
10 Stamford Road, Oakleigh, Melbourne 3166, Australia

First published 1998

Printed in the United Kingdom by Scotprint Ltd, Musselburgh

A catalogue record for this book is available from the British Library

ISBN 0 521 655560 paperback

Cover Illustration by Graham Round
Cartoons by Tim Sell